Lessons from a Population Bottleneck
Illustrated Science Exploration by Rolf A. F. Witzsche

© Text Copyright Rolf A. F. Witzsche 2018
all rights reserved

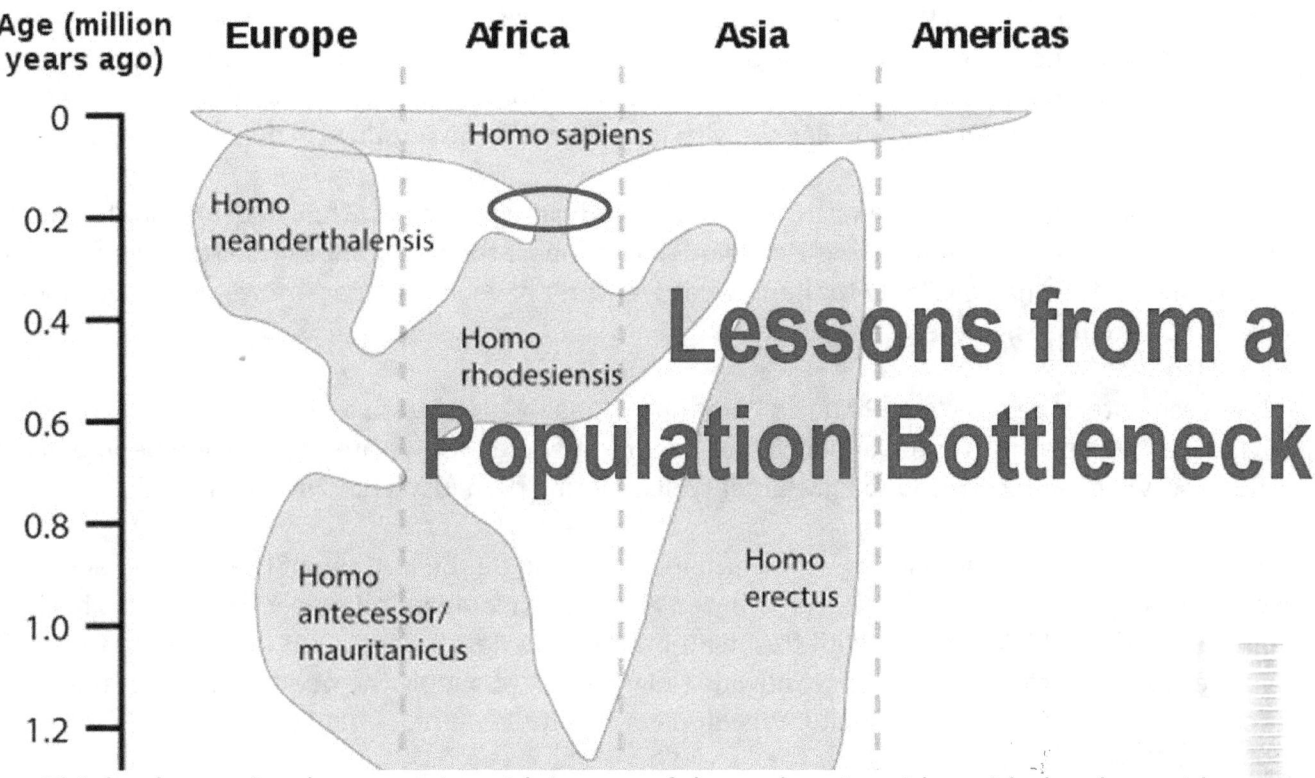

Lessons from a Population Bottleneck

This book contains the transcript with images of the exploration video with the above title:
see: http://www.ice-age-ahead-iaa.ca/

Lead in:

Living is challenging at the best of times under a 70% less-radiating Sun.
Around 195,000 years ago, the harsh climate conditions of the Ice Age began to deteriorate.
The world population was choked from 10,000s to just a few hundred that became modern humanity.

It started around 195,000 years ago, when the Ice Age climate began to deteriorate. An extended extra-deep glaciation stage began. This stage is referred to in archeology as the Marine Isotope Stage 6, or MIS6 for short.

While most of Africa became a desert at this stage and thereby became un-survivable, a small group of people who had lived at the southern tip of Africa, had evidently survived, since modern humanity is their offspring. The small group had survived in caves at a place termed Pinnacle Point. They survived, living of the sea. A warm ocean current from the more tropical northern Atlantic would have brought enough warms to the region for precipitation to occur and also some sparse vegetation to grow. A few hundred people may have lived in this region as the result of the up-lifted climate. They were the sole survivors of tens of thousands who were known to have lived in earlier times, who perished for the lack of resources that their primitive cultures had been unable to supply, in spite of the growing intelligence of humanity.

Population-collapse events of this type where the world population falls back to minuscule levels in the shadow of collapsing climates, are referred to in archeology as Population Bottleneck events. We are facing another population bottleneck of gigantic proportions in the near future. People need food and water to live. Without them they die. We are on the fast track of loosing both food and water. A sharp down-ramping of the solar activity has begun. The Earth is getting colder and drier. The climate that gave us the power to support us as a seven billion world population, is collapsing fast, globally. And as the climate collapses, the food supply collapses with it, and with it the population density collapses.

In this regard, the people at Pinnacle Point had an advantage over us. They had lived in the one area in which the natural environment was rich enough to support their existence under the worst possible circumstances. They had available to them, what we won't have in the near future, which we have the power to create for us, but refuse to do so.

For freshwater, the people at Pinnacle Point had probably discovered a well or spring nearby that had provided enough freshwater to support a few-hundred people. But we are seven billon people in the world now, we need a correspondingly larger well or spring than just a hold in the ground.

The equivalent of the well-spring at Pinnacle Point, would be the outflow of the Amazon and Congo Rivers, and so on, on the world-scale, distributed globally. It would be relatively easy, on this basis, to supply a world of 7 billion people with freshwater during the desert conditions in glaciation times. We have the resources to build this big, worldwide - big enough to supply the needs of industries and agricultures for the whole of humanity. Why don't we do it?

Table of Contents

The development of life on Earth began more than 500 million years ago 7

The breakout of humanity roughly 2 1/2 million years ago... 8

During the Ice Age epochs the Sun operates in low-power mode 9

Cosmic rays pass right through our body... 10

As they pass through us, they generate an electric current by induction 11

Humanity became a resourceful intelligent species in the shadow large volumes of cosmic ray-flux 12

We became so successful in our self-development.. 13

Living is challenging under a 70% less-radiating Sun 14

Precipitation is dramatically less during glaciation conditions 15

Around 195,000 years ago extended extra-deep glaciation began... 16

The small group had survived in caves at Pinnacle Point 17

The population collapse became of immense magnitude.. 18

To avoid a climate-caused population collapse crisis 19

This means in modern time, because agriculture is the main-food resource 20

The people who hadn't lived at Pinnacle Point 21

We have the technological capacity to build us a whole new world.. 22

We are presently progressing towards the next Ice Age phase shift 23

A sharp down-ramping of the solar activity has begun 24

The people at Pinnacle Point had probably discovered a spring... 25

We need a wellspring big enough for the global scale 26

Freshwater is critical for human living, at any time.. 27

While the needed water is plentifully available.. 28

The outflow of the Columbia River could solve the California water crisis 29

We face the same irony also on the global scale 30

The people at Pinnacle Point also had another advantage 31

Modern society has committed itself to become aimless refugees 32

The few hundred people at Pinnacle Point had lived largely of the sea 33

Only large-scale high-tech indoors agriculture has the capacity 34

That's why the people at Pinnacle Point survived 35

No provisions that would assure our continued living 36

Humanity has the power to build for itself the New World with joy 37

More Illustrated Science Books by Rolf A. F. Witzsche 38

The development of life on Earth began more than 500 million years ago

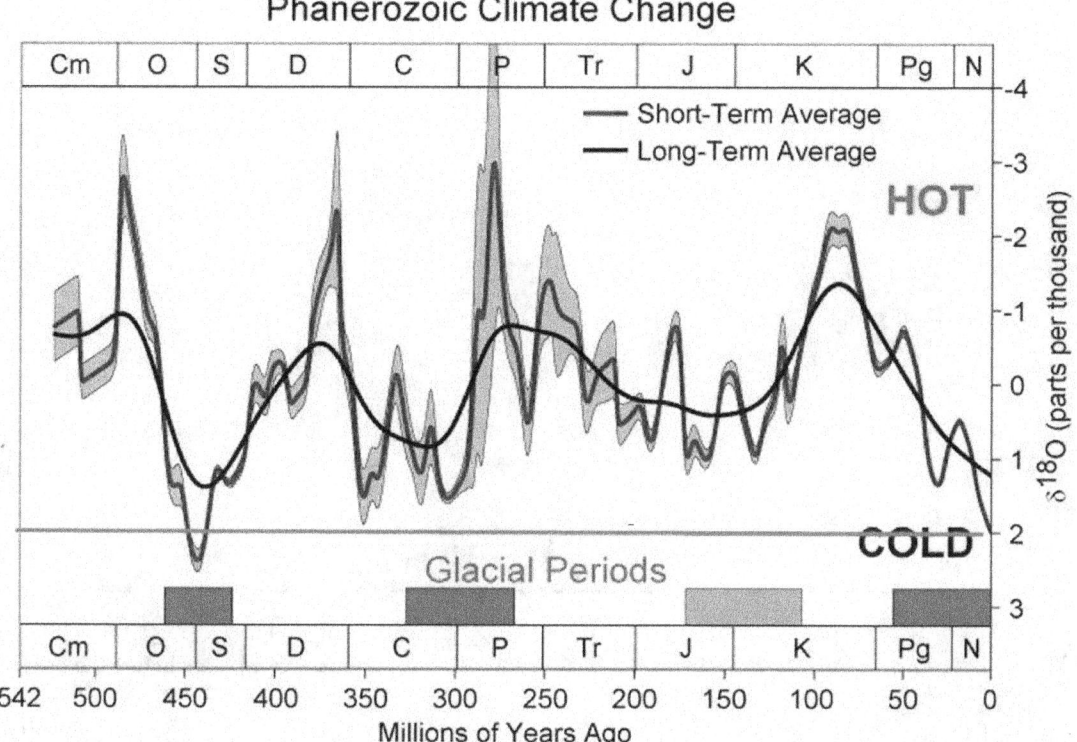

The development of life on Earth began more than 500 million years ago. It took almost this entire long period, for life to advance far enough for the dawn of humanity to occur.

The breakout of humanity roughly 2 1/2 million years ago

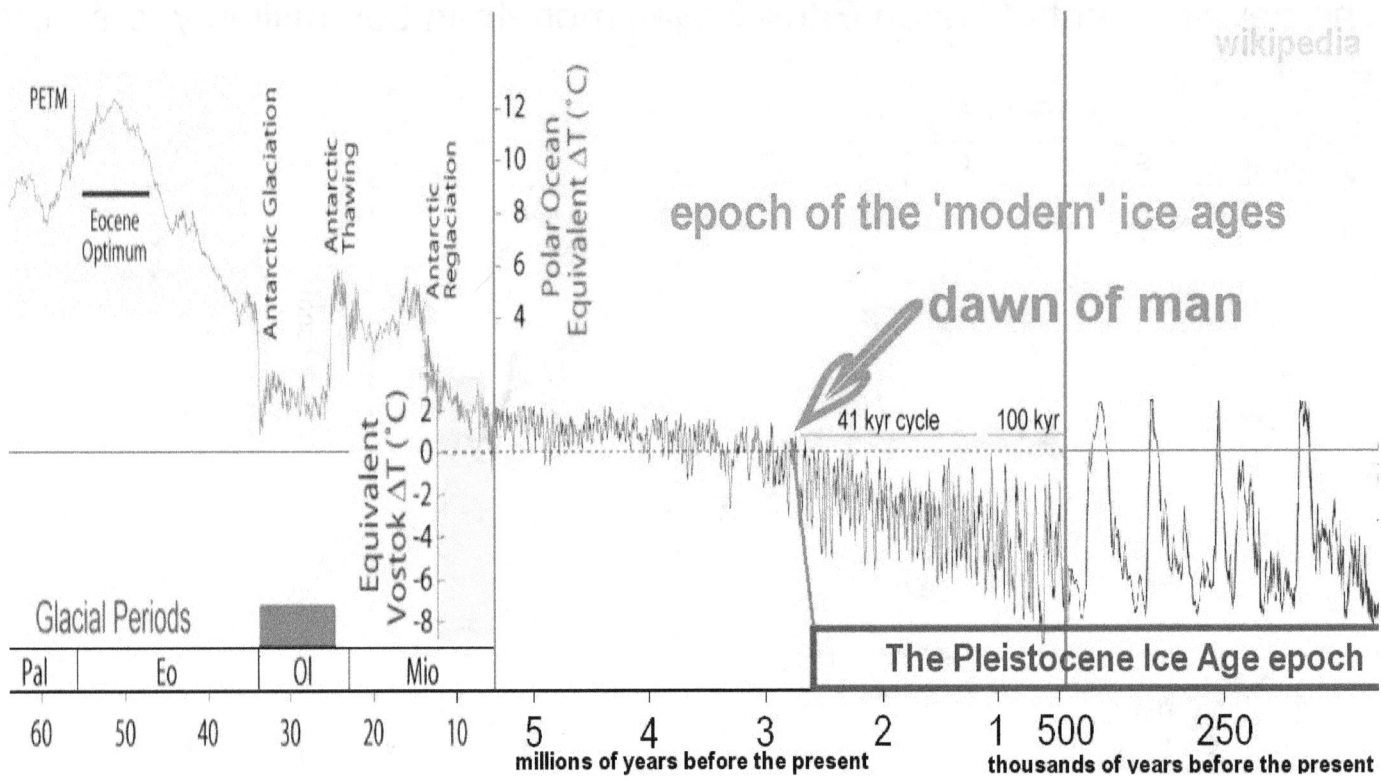

The breakout of humanity into becoming the high-order intelligent species that it became, coincided with the beginning of the modern Ice Age epoch roughly 2 1/2 million years ago.

During the Ice Age epochs the Sun operates in low-power mode

During the Ice Age epochs the Sun operates in low-power mode, with 70% less energy being radiated by it, while it radiates significantly larger volumes of cosmic-ray flux.

Cosmic rays pass right through our body

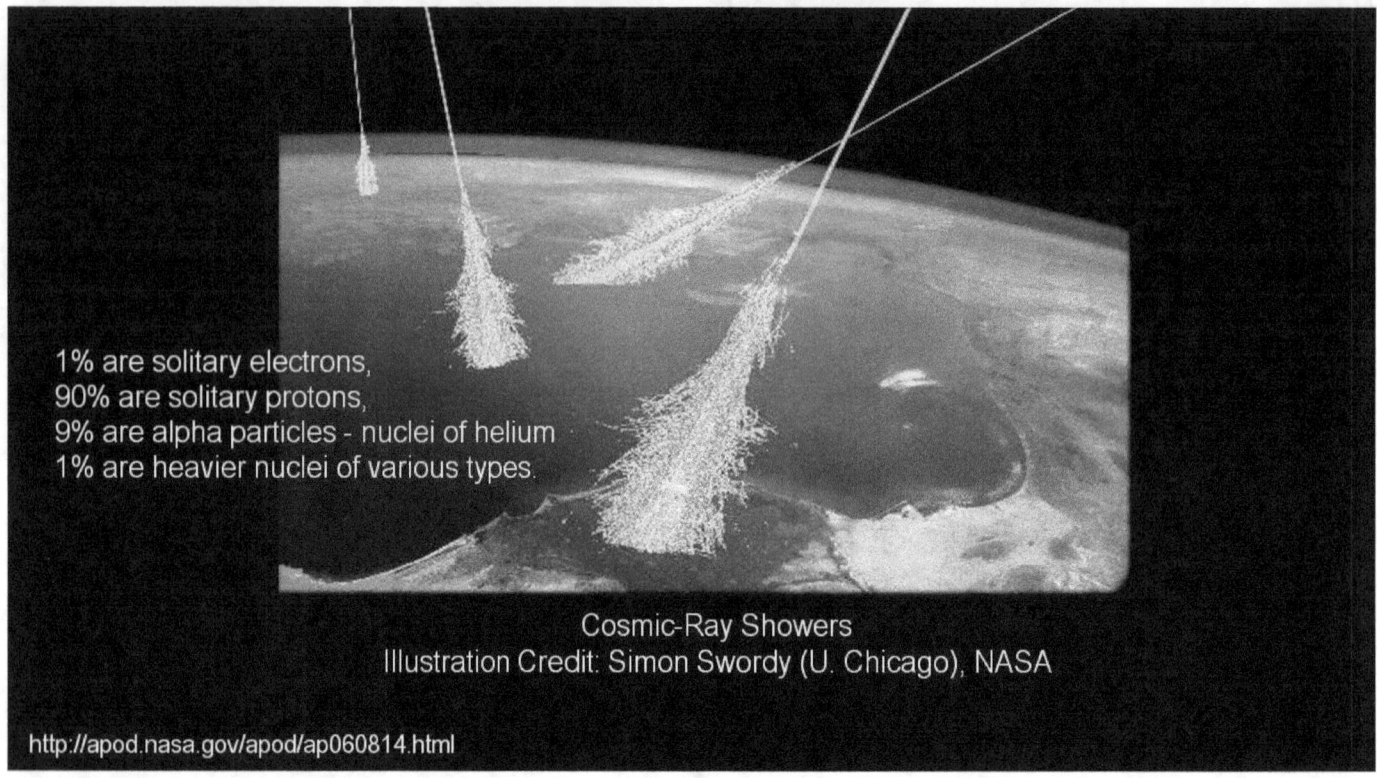

Cosmic rays are single events of highly energized electrically charged particles. Most cosmic-ray flux originates at the Sun. Some events pass clean through our atmosphere, and some even pass right through our body.

As they pass through us, they generate an electric current by induction

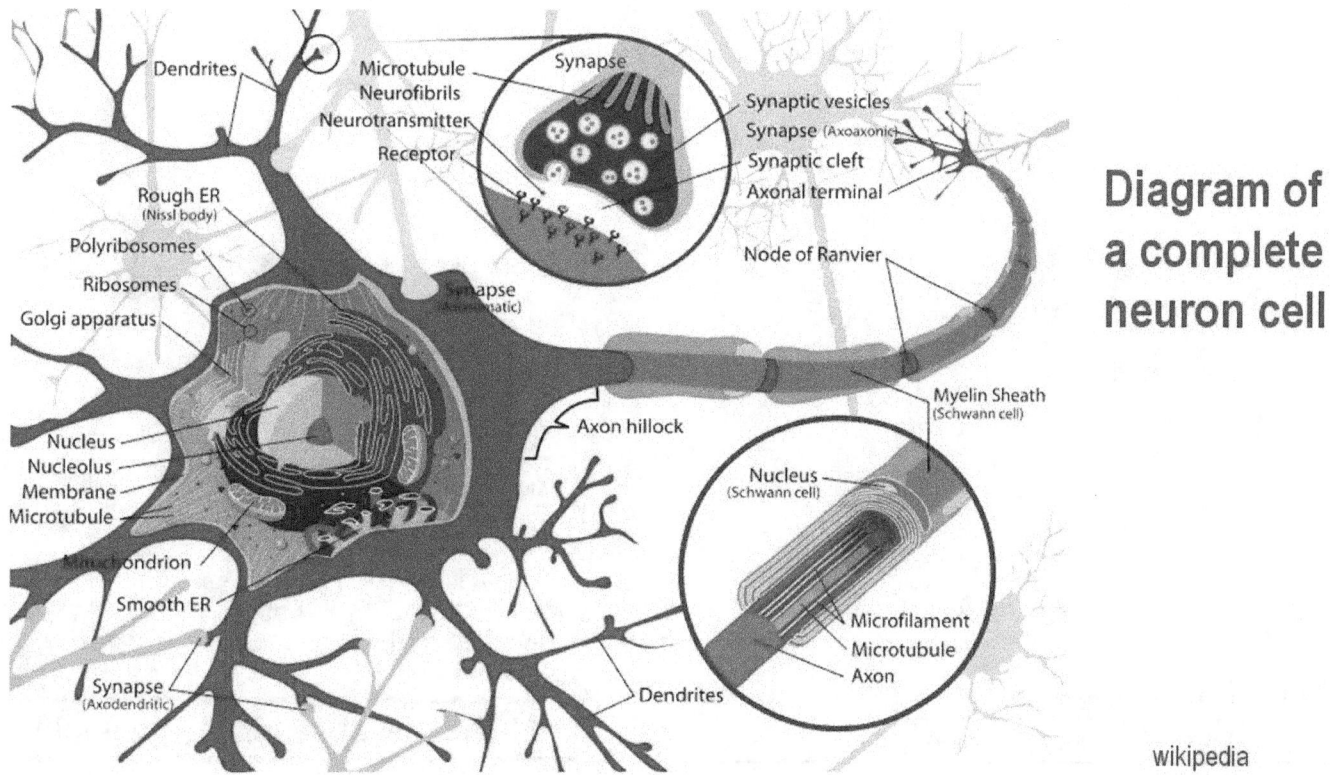

Diagram of a complete neuron cell

wikipedia

As they pass through us, they generate an electric current by induction, which appears to enhance our neurological systems, that operates to some degree eclectically, including our cognitive systems.

Humanity became a resourceful intelligent species in the shadow large volumes of cosmic ray-flux

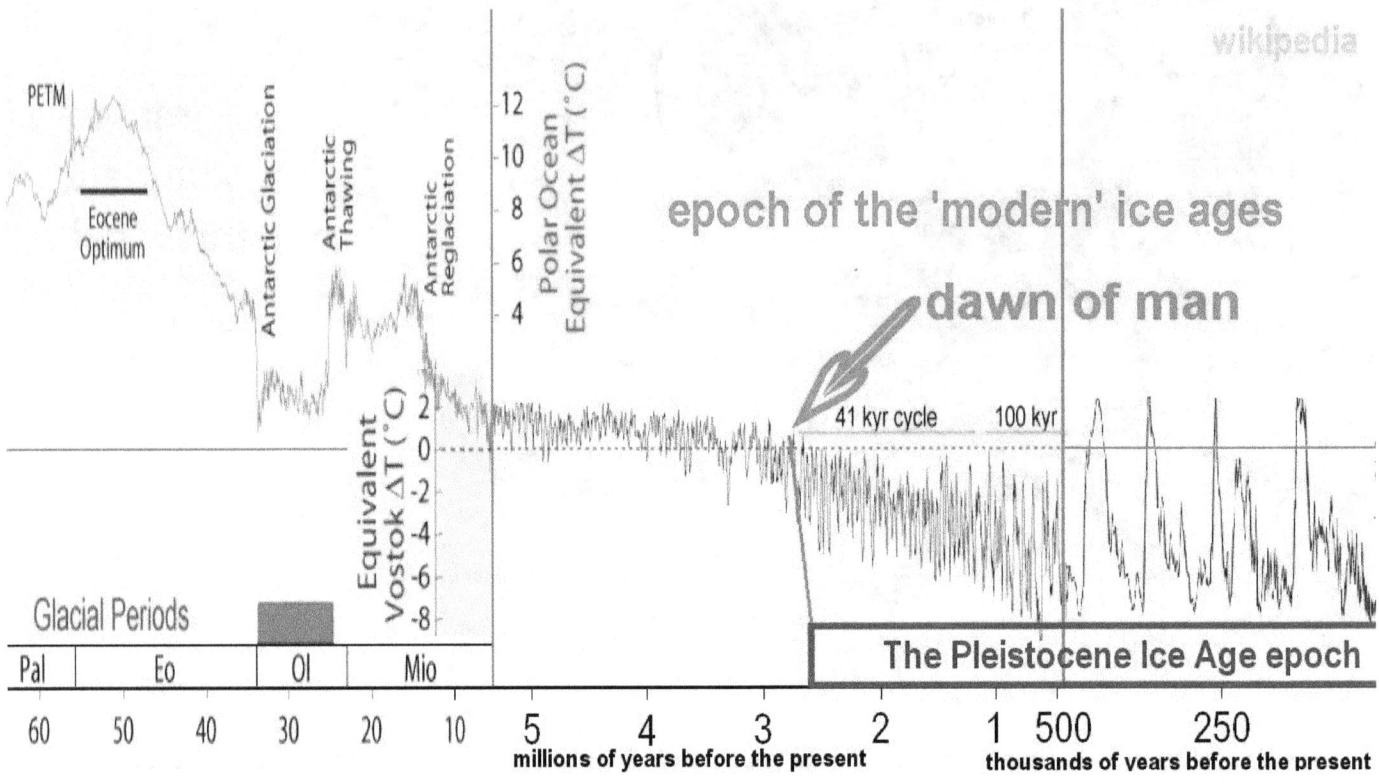

So it was that humanity became a resourceful intelligent species in the shadow of 2 1/2 million years of cyclical ice ages with large volumes of cosmic ray-flux.

We became so successful in our self-development

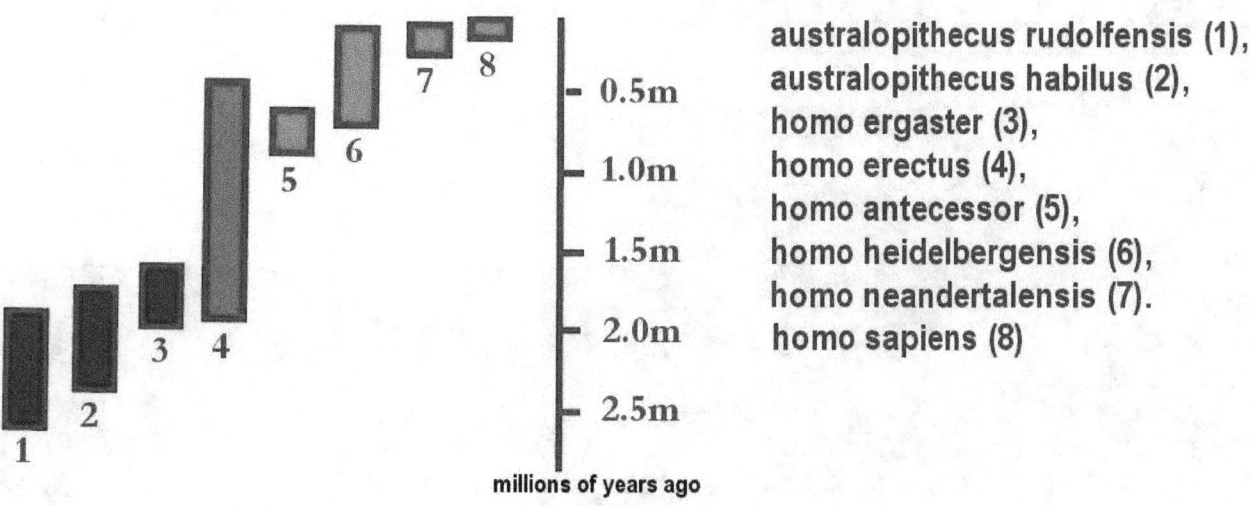

We, the homo sapiens (8), are the only surviving, and the shortest lived of all the the human species, at barely 200,000 years of age.

We became so successful in our self-development on this course that we proliferated into 8 distinct human species, of which we are the last one, and the sole survivor.

The reason why we are the sole survivor is rooted in the nature of the Ice Age phenomenon.

Living is challenging under a 70% less-radiating Sun

Living is challenging at the best of times under a 70% less-radiating Sun. The climate is cold and dry during the glaciation periods.

Precipitation is dramatically less during glaciation conditions

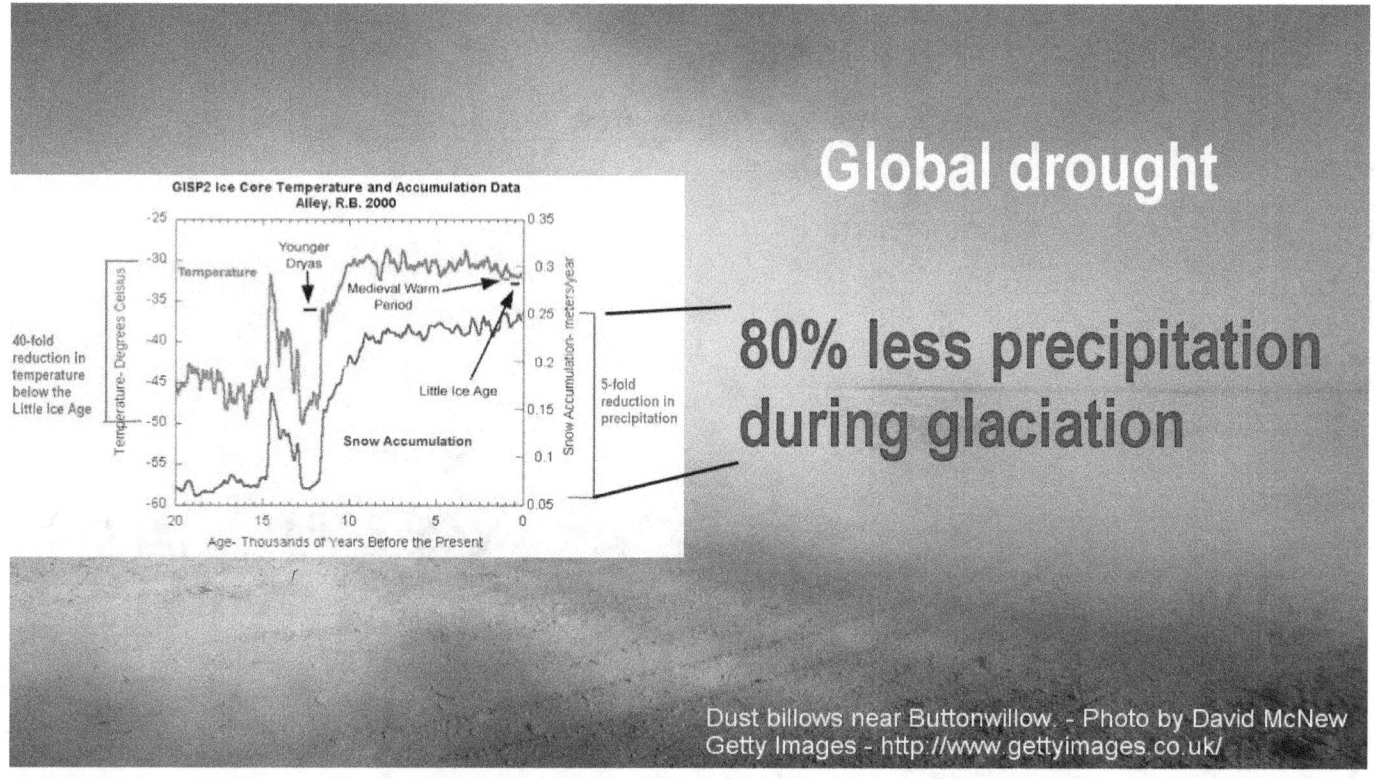

We also know from ice core examinations that precipitation is dramatically less during glaciation conditions, as much as 80% less. Plant growth becomes devastated by this double effect of cold and dry condition, while plant growth is what all life depends on.

In extreme times of such condition, humanity becomes overwhelmed by the conditions that it can no longer live with. Without food, people die. Without water, people die.

Around 195,000 years ago extended extra-deep glaciation began

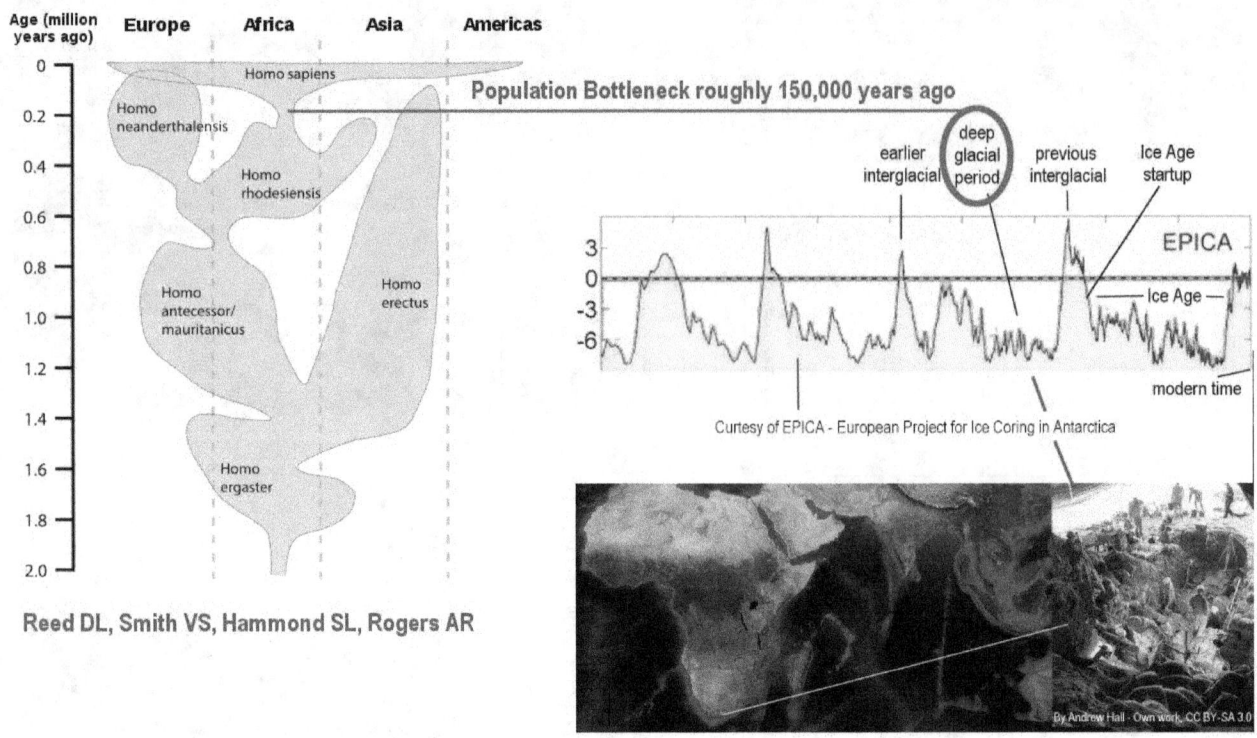

An extremely harsh case of this type occurred during the second-last Ice Age. It started around 195,000 years ago, when the already harsh climate conditions of the Ice Age began to deteriorate. An extended extra-deep glaciation stage began. This stage is referred to in archeology as the Marine Isotope Stage 6, or MIS6 for short.

While most of Africa became a desert at this stage and thereby became un-survivable, a small group of people who had lived at the southern tip of Africa, had evidently survived, since modern humanity is their offspring.

The small group had survived in caves at Pinnacle Point

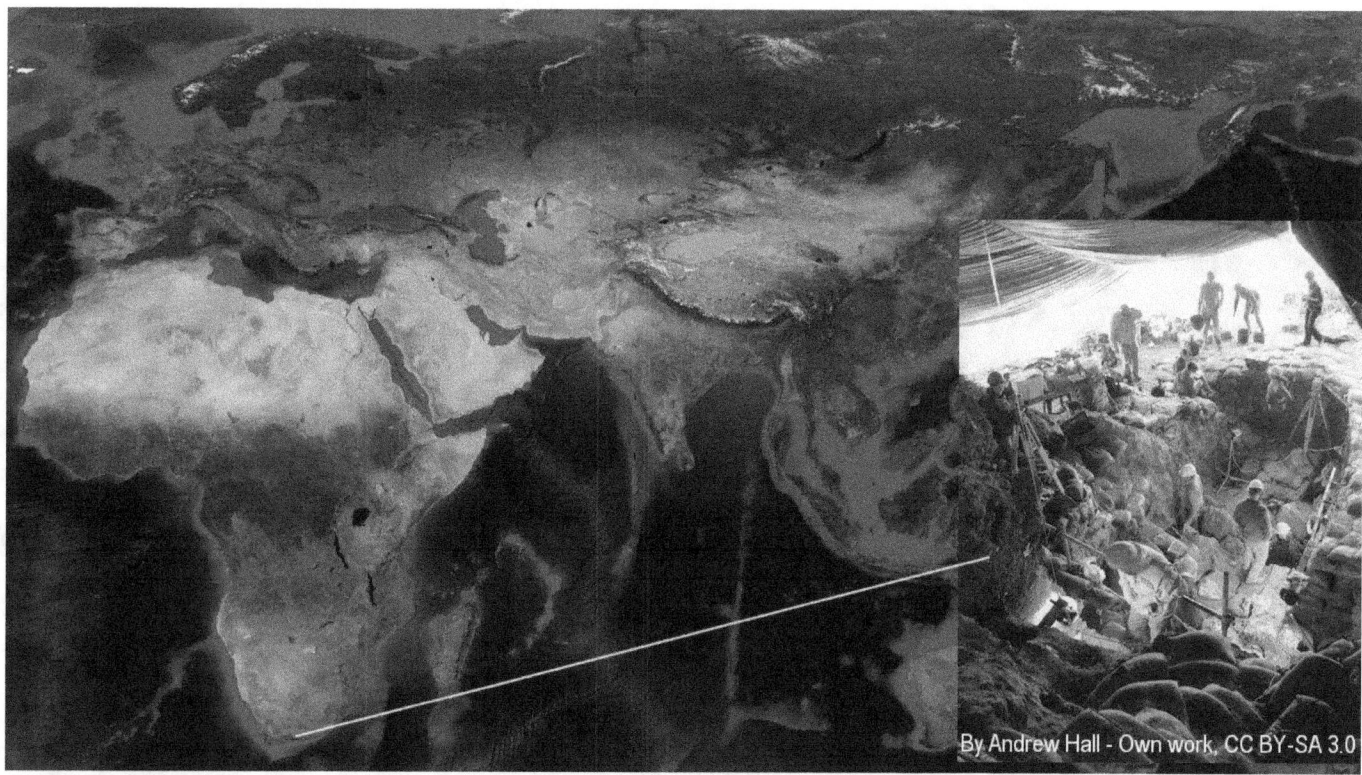

The small group had survived in caves at a place termed Pinnacle Point. They survived, living of the sea. A warm ocean current from the more tropical northern Atlantic would have brought enough warms to the region for precipitation to occur and also some sparse vegetation to grow. A few hundred people may have lived in this region as the result of the up-lifted climate. They were the sole survivors of tens of thousands who were known to have lived in earlier times, who perished for the lack of resources that their primitive cultures had been unable to supply, in spite of the growing intelligence of humanity.

The selection of caves at Pinnacle Point for their living might have been the leading edge of intelligent reasoning that humanity had developed at the time. It saved their existence.

The population collapse became of immense magnitude

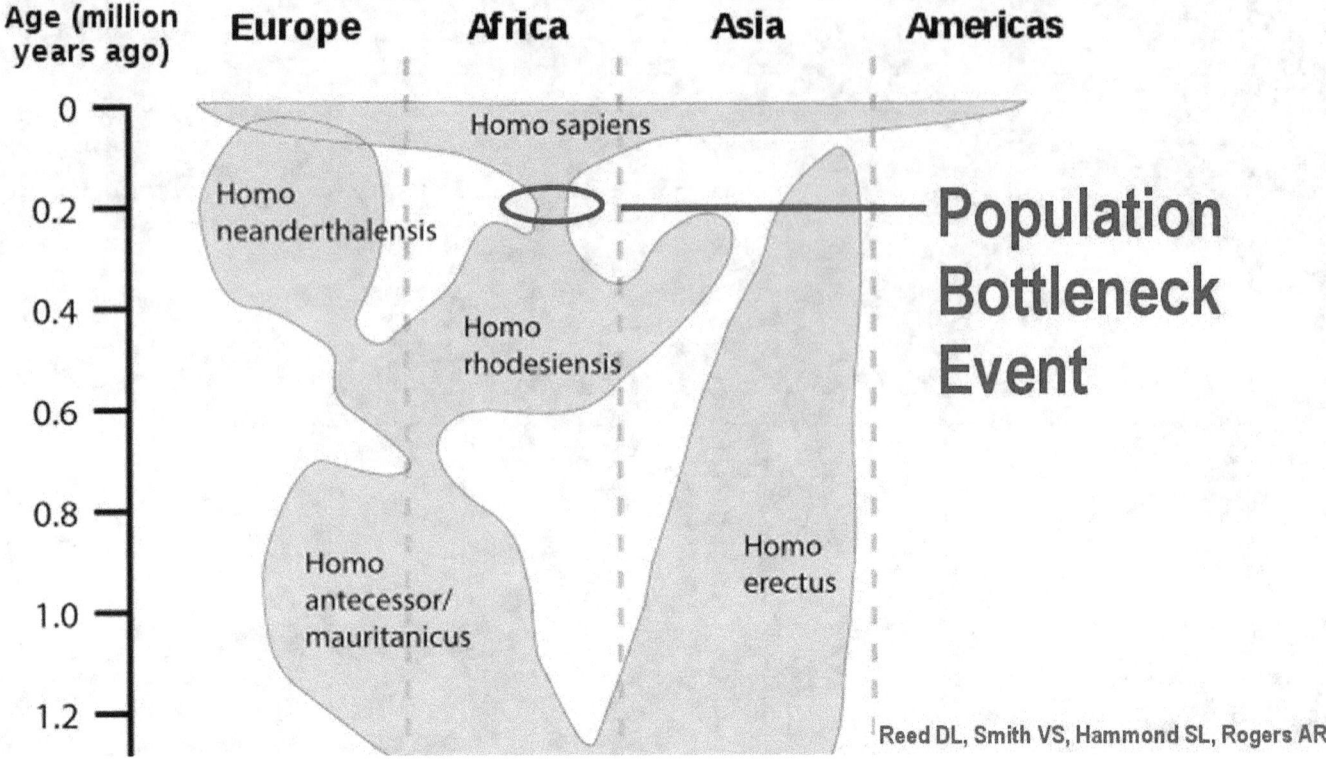

Population Bottleneck Event

The population collapse became of immense magnitude that the other human cultures had failed to avoid, at this time. Population-collapse events of this type, where the world population falls back to minuscule levels in the shadow of collapsing climates, are referred to in archeology as Population Bottleneck events.

The particular bottleneck event that the people at Pinnacle Point had survived, likely had occurred in the general timeframe at around 150,000 years ago.

To avoid a climate-caused population collapse crisis

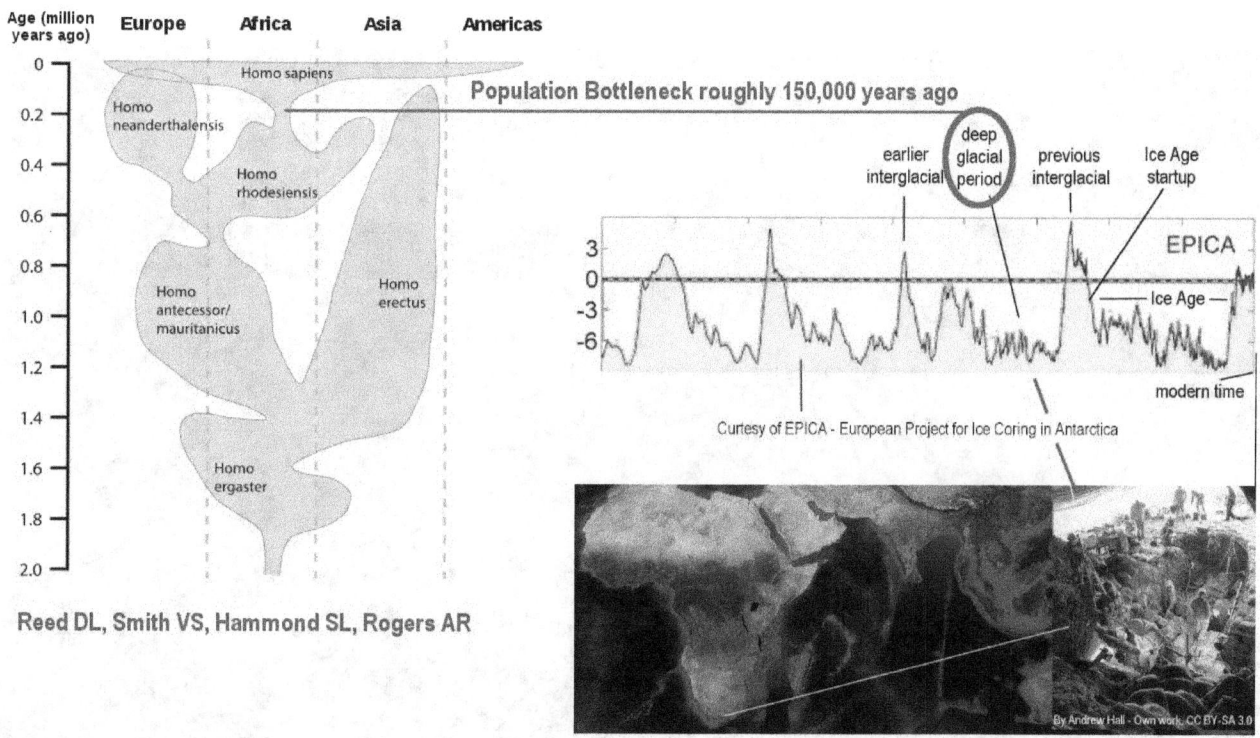

The massive population collapse during the bottleneck period illustrates rather dramatically the immensely devastating impact that a major climate collapse can have, which even the most intelligent species could not survive, which humanity evidently had been at the time.

This means that when the climate dies, people die with it.

The only option that exists under such circumstances, to avoid a climate-caused population collapse crisis, is to transfer the food-production infrastructure out of the natural environment, into a technological environment that the climate collapse cannot affect.

This means in modern time, because agriculture is the main-food resource

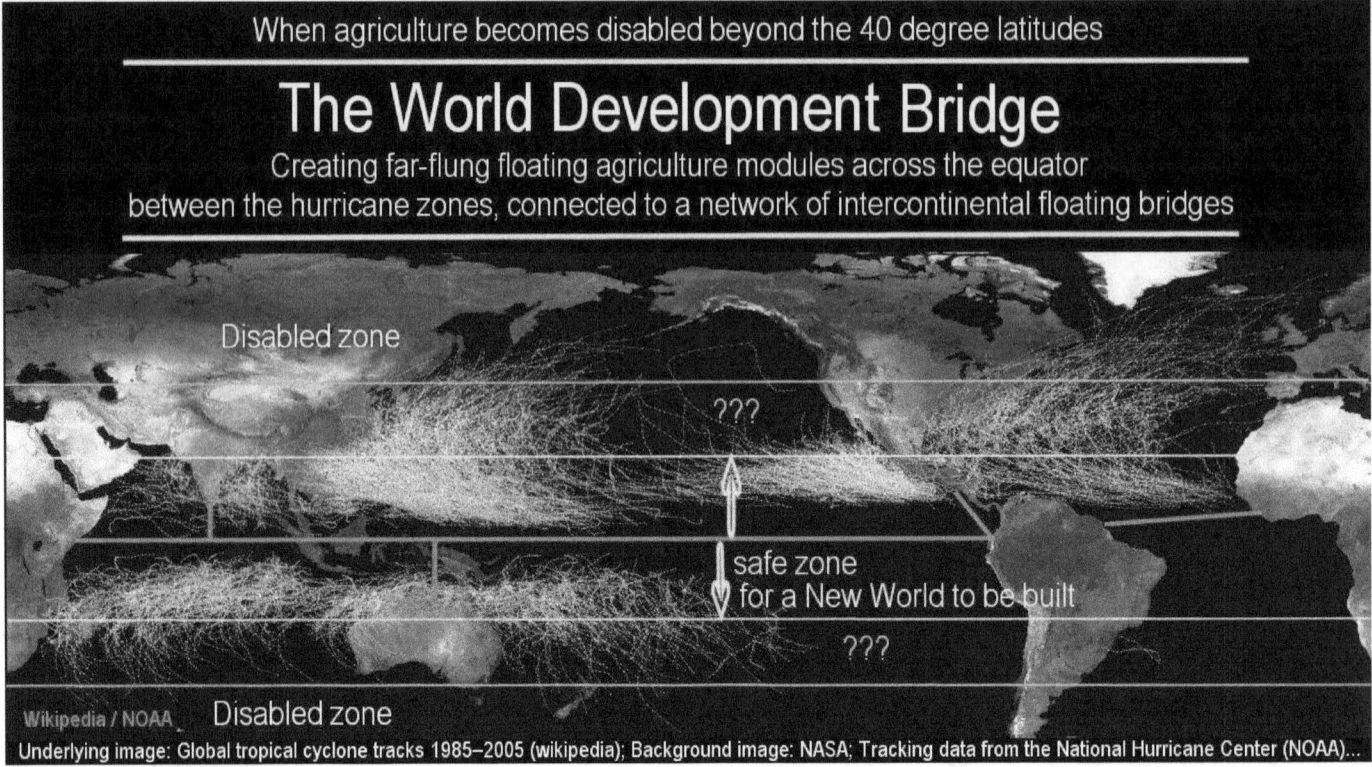

This means in modern time, because agriculture is the main-food resource for humanity by far, that agriculture becomes placed into indoor facilities and into regions of the planet that are physically secure from climate calamities.

The people who hadn't lived at Pinnacle Point

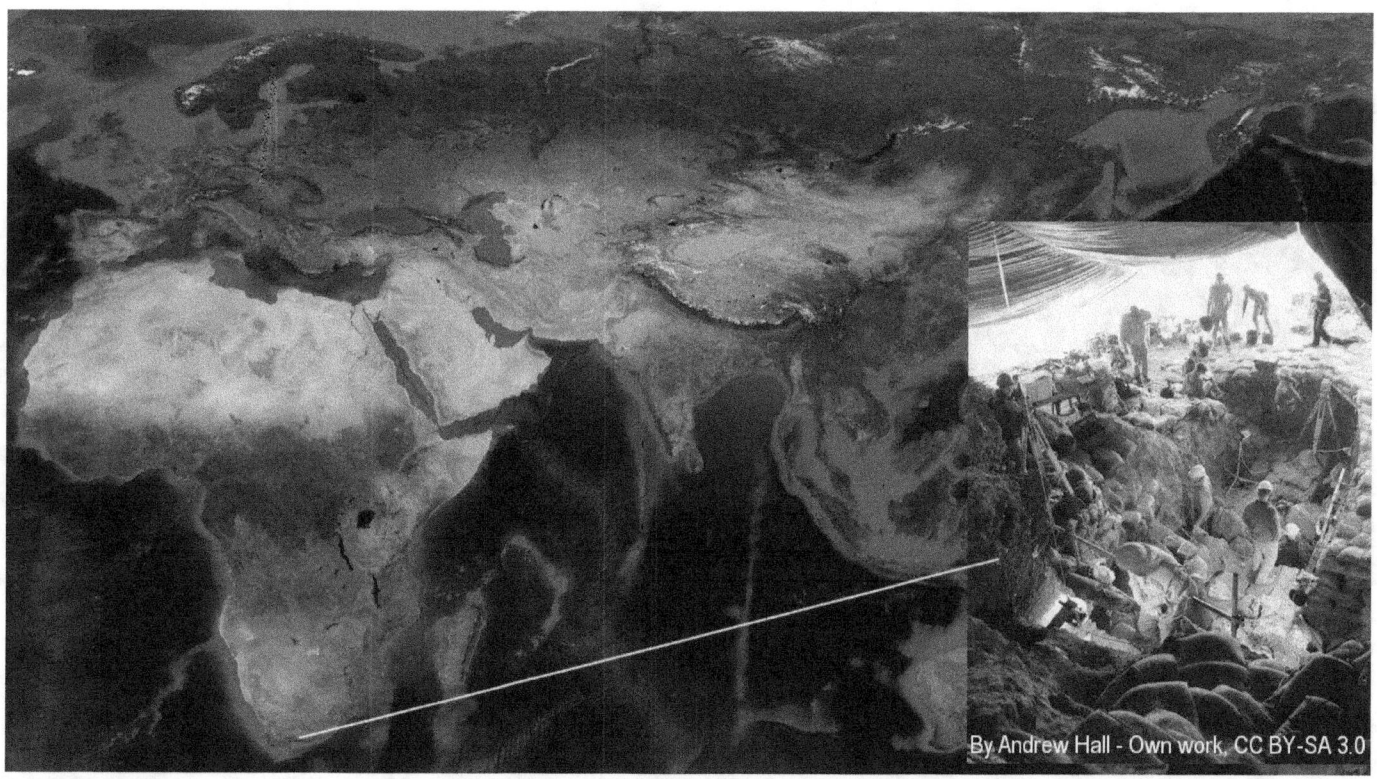

The people who hadn't lived at Pinnacle Point, during the bottleneck population collapse 150,000 years ago, didn't have this option open to them, to build themselves a new world. They lacked the capacity to do so. Consequently they perished. All but a few hundred of the entire population of humanity perished for this type of reason.

We have the technological capacity to build us a whole new world

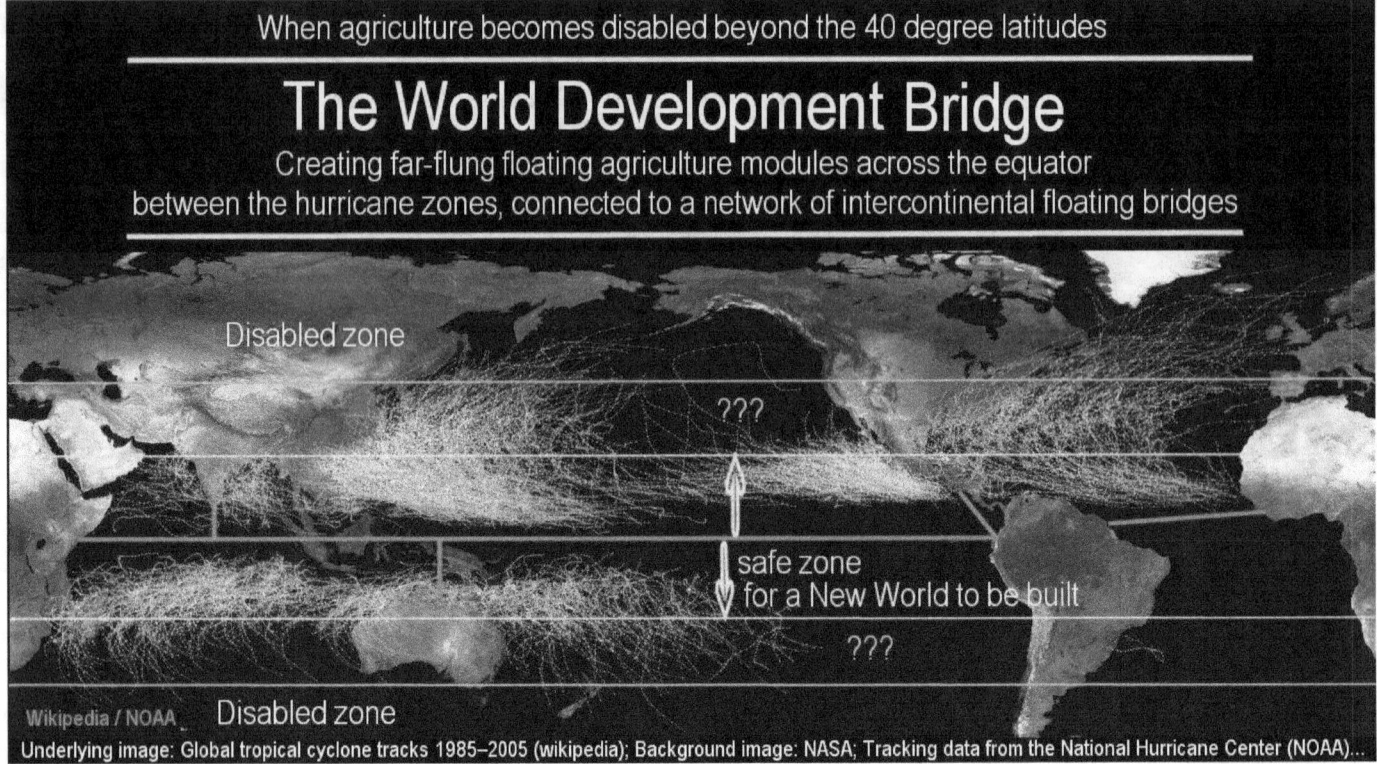

But we in modern time, do have this option open to us today to build us a new world. We have the technological capacity to build us a whole new world that the Ice Age climate cannot affect. With this capacity we have the power at hand to avoid the greatest potential population-collapse crisis that appears otherwise assured, which may be upon us in the 2050s.

We are presently progressing towards the next Ice Age phase shift

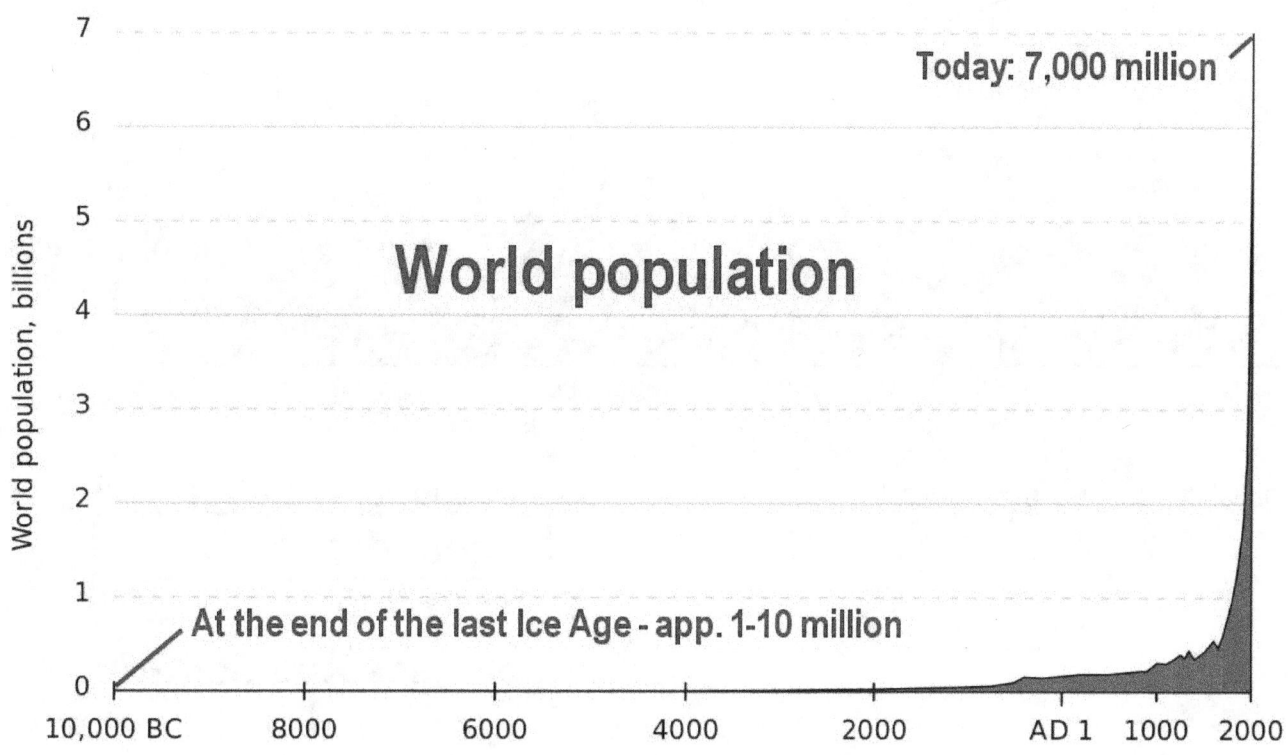

We are presently progressing towards the next Ice Age phase shift with a seven-billion world population, of which few would get past the population bottleneck. The reason is simple.

A sharp down-ramping of the solar activity has begun

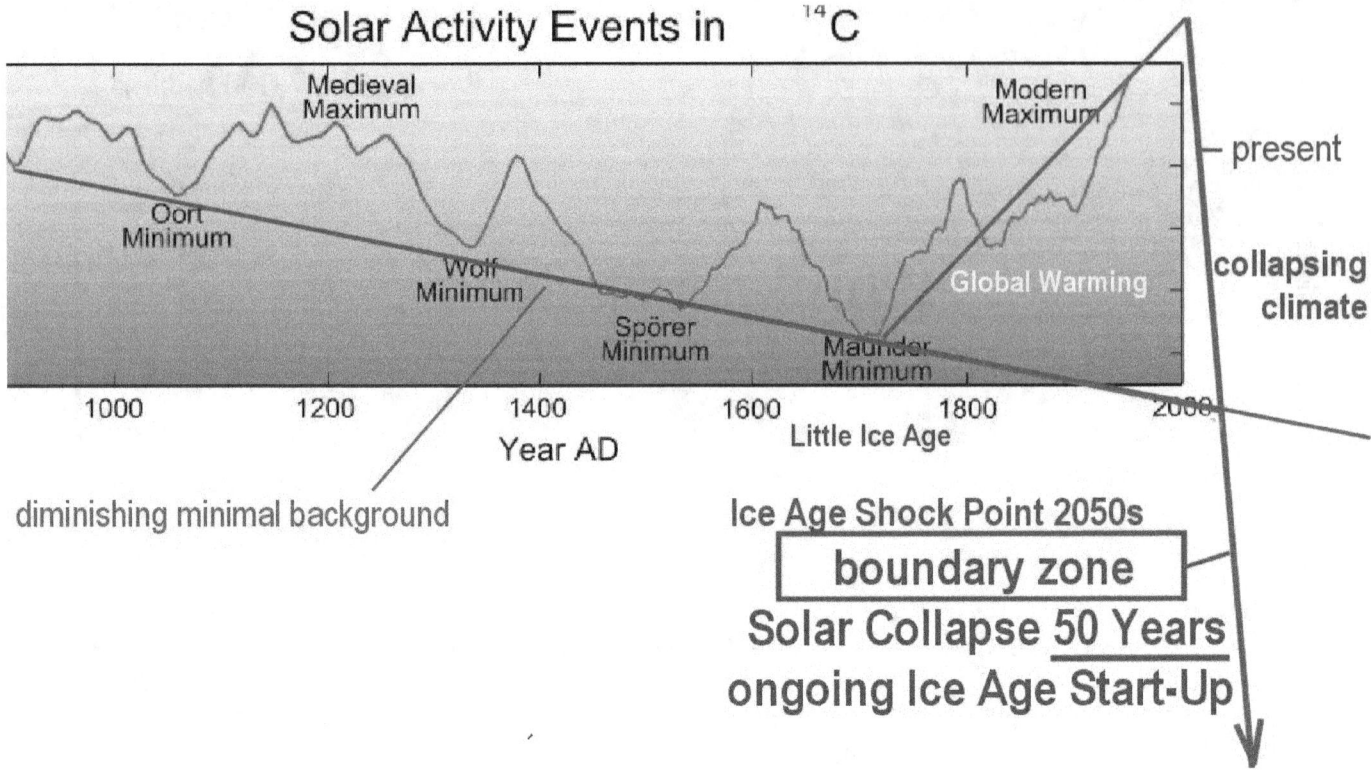

A sharp down-ramping of the solar activity has begun. The Earth is getting colder and drier. The climate that gave us the power to support us as a seven billion world population, is collapsing fast, globally. And as the climate collapses, the food supply collapses with it, and with it too, the population density collapses. We are facing another population bottleneck of gigantic proportions. People need food and water to live. Without them they die. We are on the fast track of loosing both food and water.

In this regard, the people at Pinnacle Point had an advantage over us. They had lived in the one area in which the natural environment was rich enough to support their existence, and this under the worst possible circumstances. Their place may have been the only place in the world where this was possible, because all of the rest of humanity died out.

The few surviving people had the advantage over us in that they had available to them, what we won't have in the near future, which we have the power to create, but refuse to apply that power.

The people at Pinnacle Point had probably discovered a spring

For freshwater, the people at Pinnacle Point had probably discovered a spring of groundwater nearby that had provided enough freshwater to support a few-hundred people.

But we are seven billon people in the world now, we need a correspondingly larger wellspring than just a hole in the ground.

We need a wellspring big enough for the global scale

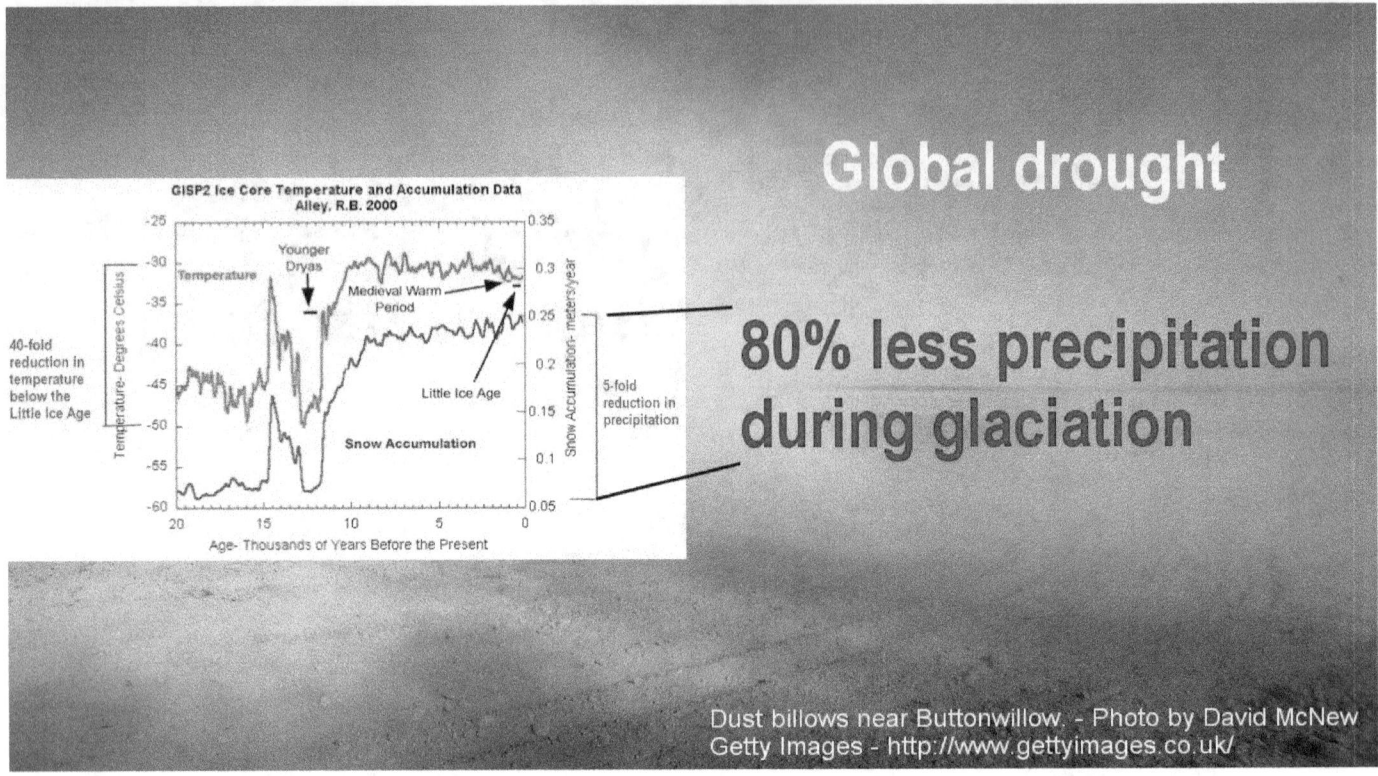

We need a wellspring big enough for the global scale, and we need it when precipitation diminishes by 80% worldwide.

In fact, we should have it already started, because drought conditions are increasing in many placed of the world.

Freshwater is critical for human living, at any time

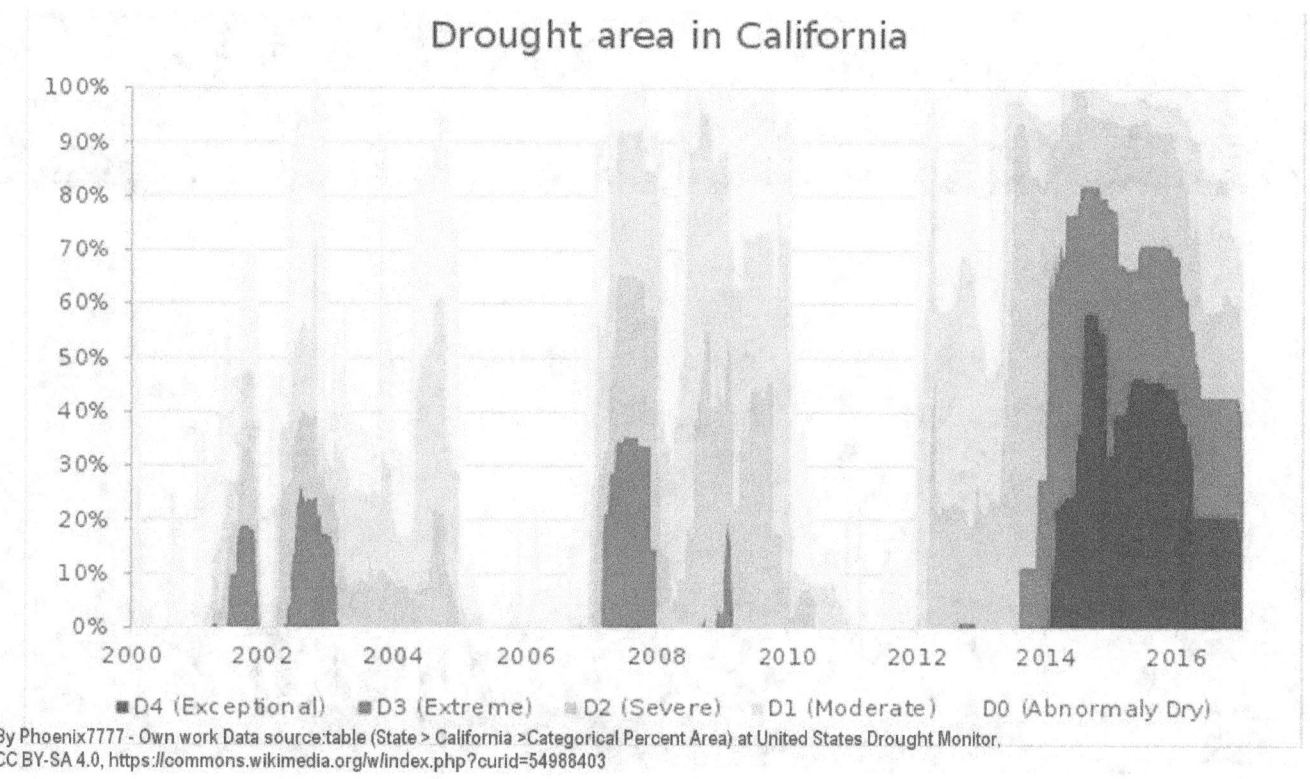

Freshwater is critical for human living, at any time, and any place, and for all life. This fact is already felt in California. The big drought began in 2010. Agriculture is severely impacted.

While the needed water is plentifully available

While the needed water is plentifully available nearly, nothing is being done to get it to where it is needed.

The outflow of the Columbia River could solve the California water crisis

The outflow of the Columbia River, for example, could solve the California water crisis easily, if the water was conveyed to California with the technologies that we have available.

The the technological diversion of the outflow of the Columbia River within a large-scale artery floating is the sea, would be for California the equivalent of the people at Pinnacle Point finding a well-spring nearby, or building one.

The irony is, that we have this capability at hand in the modern world, nothing is done along this line to fulfill the human need, nor does anyone care to cause the needed steps to happen.

We face the same irony also on the global scale

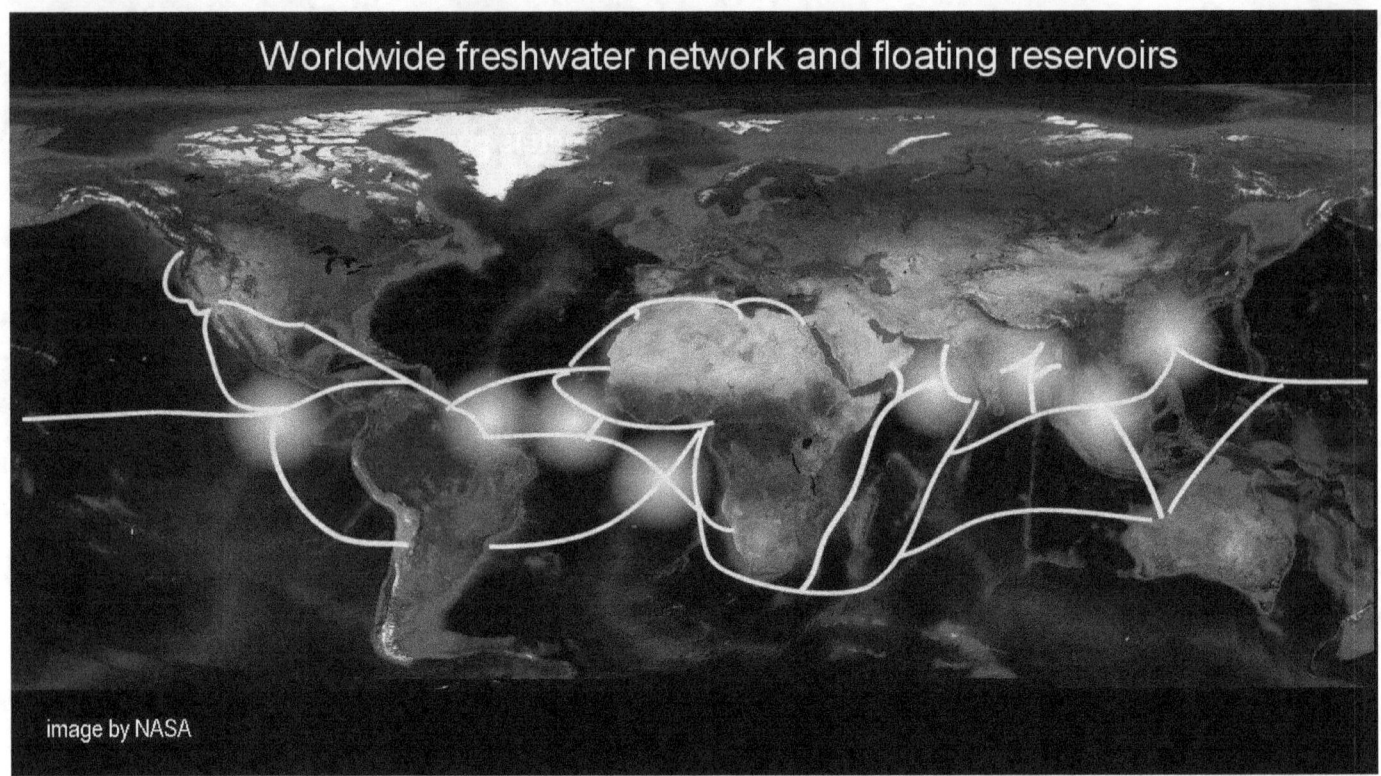

We face the same irony also on the global scale. We know that the phase shift to the Ice Age will collapse rainfall by 80%, but we refuse to make the preparations for it. The process would be the same in principle as bringing water to California from the Columbia River. The source would have to be larger, of course, to meet the global need in an Ice Age environment.

The equivalent of the well-spring at Pinnacle Point, would be the outflow of the Amazon and Congo Rivers, and so on, on the world-scale, distributed globally. It would be relatively easy, on this basis, to supply a world of 7 billion people with freshwater during the desert conditions in glaciation times. So, why isn't this done? We have the resources to build this big, worldwide - big enough to supply the needs of industries and agricultures for the whole of humanity. Why don't we do it?

People say that hell will freeze over before such a worldwide water project will be built.

But if it isn't built? The refusal to build this infrastructure, and to build it big, means that the giant population collapse that could be avoided, will thereby be invited.

This means that the refusal to build this infrastructure is akin to accepting the death sentence for almost the whole of humanity.

That's the context in which the building of what is needed, will likely happen, because there is no joy in being dead.

The people at Pinnacle Point also had another advantage

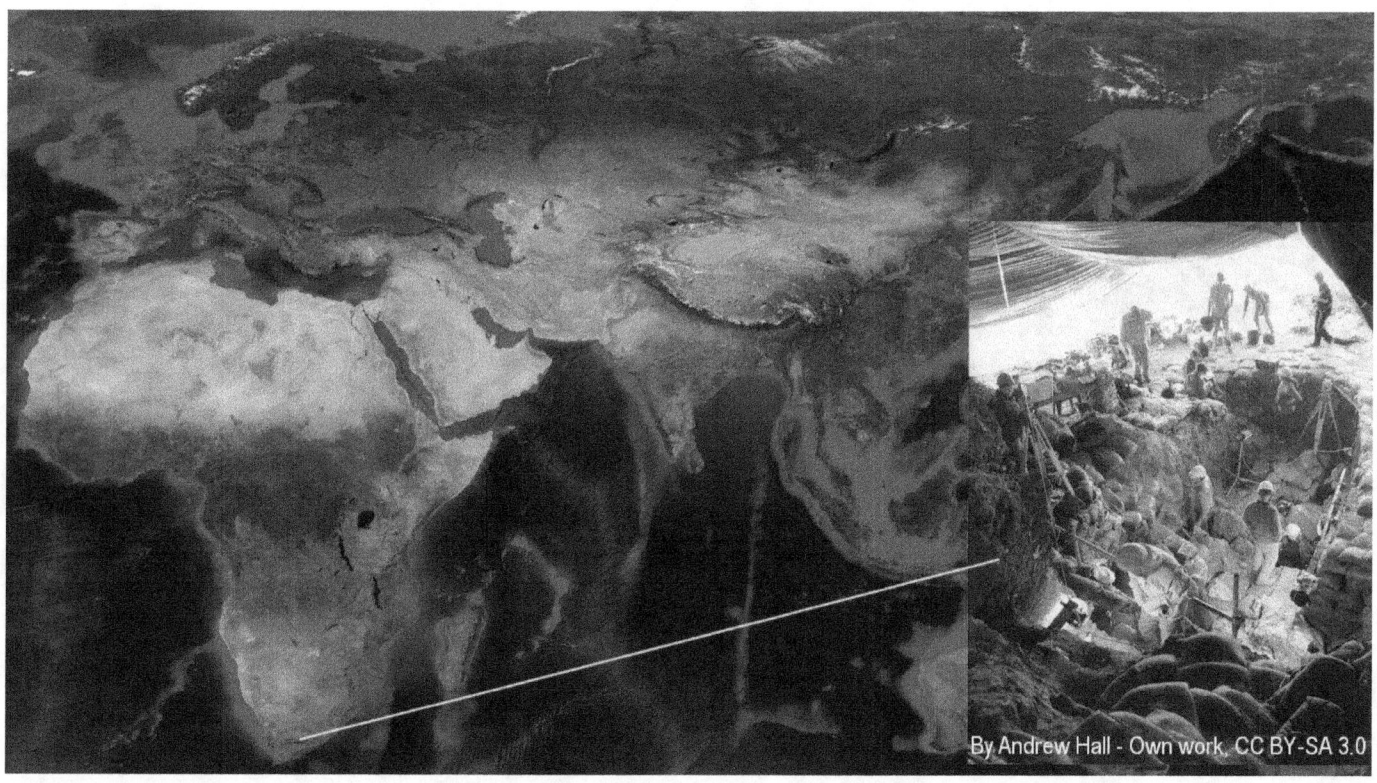

The people at Pinnacle Point also had another advantage that modern society is determined not have itself. The people had a home in which to live during the climate collapse period.

Modern society has committed itself to become aimless refugees

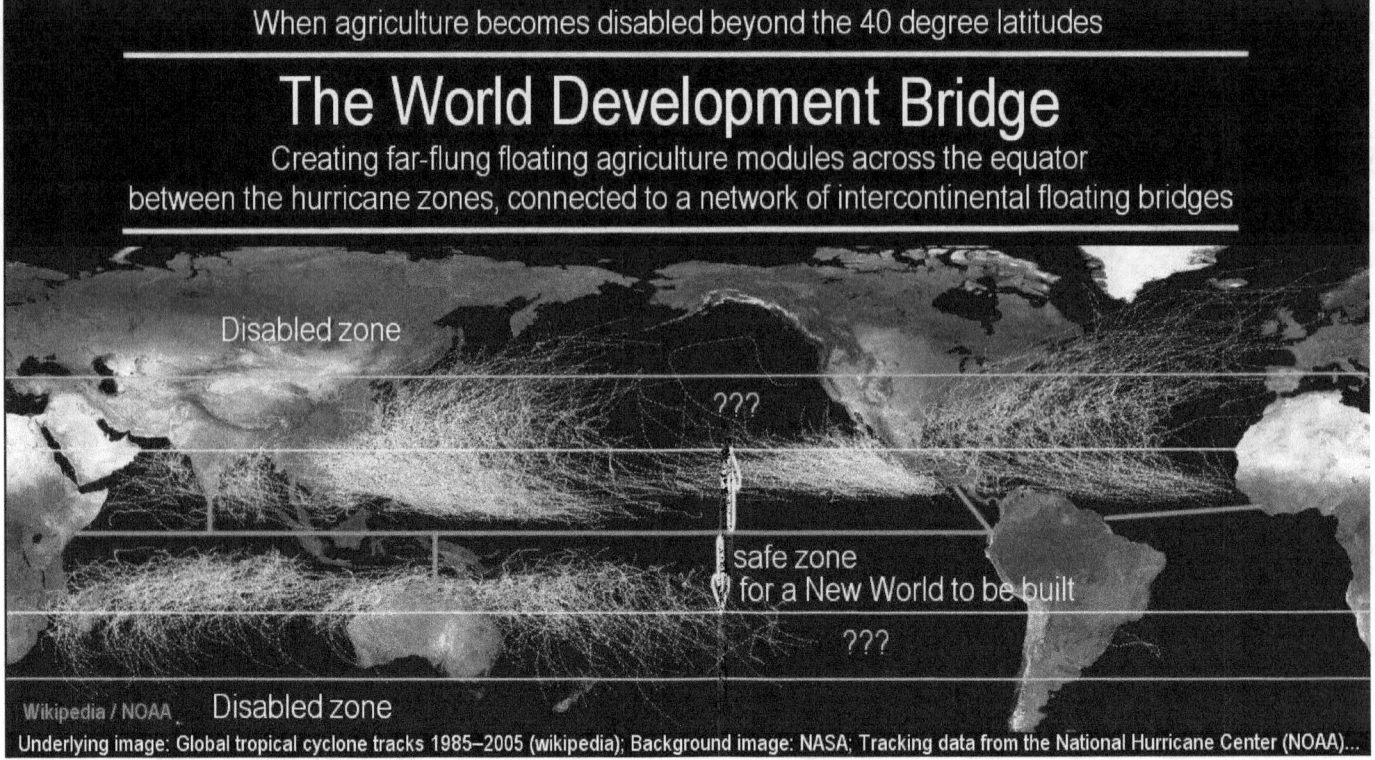

Society doesn't have the slightest intention today to provide itself the equivalent. When things got rough for the people at Pinnacle Point, they had a cave to go to. Modern society has no place prepared to go to when the northern regions from Canada to Europe to Russia become uninhabitable by the cold, or even before that when agriculture begins to fail.

Modern society has committed itself to become aimless refugees, because the building of the 6,000 new cities for society to have a home to migrate to, is not happening. It is not even considered.

The needed grand-style building project, the grandest-ever imagined, for society to live and have a future, is being laughed at today as an utopian dream.

The people at Pinnacle Point, however, would laugh at us, instead, for our smallness at heart.

The few hundred people at Pinnacle Point had lived largely of the sea

Another key-element that the people at Pinnacle Point evidently had going for them, which modern society refuses to even consider, is food.

The few hundred people at Pinnacle Point had lived largely of the sea. The sea was rich enough to support those few-hundred people, but the sea won't support seven-billion people for 90,000 years.

Only large-scale high-tech indoors agriculture has the capacity

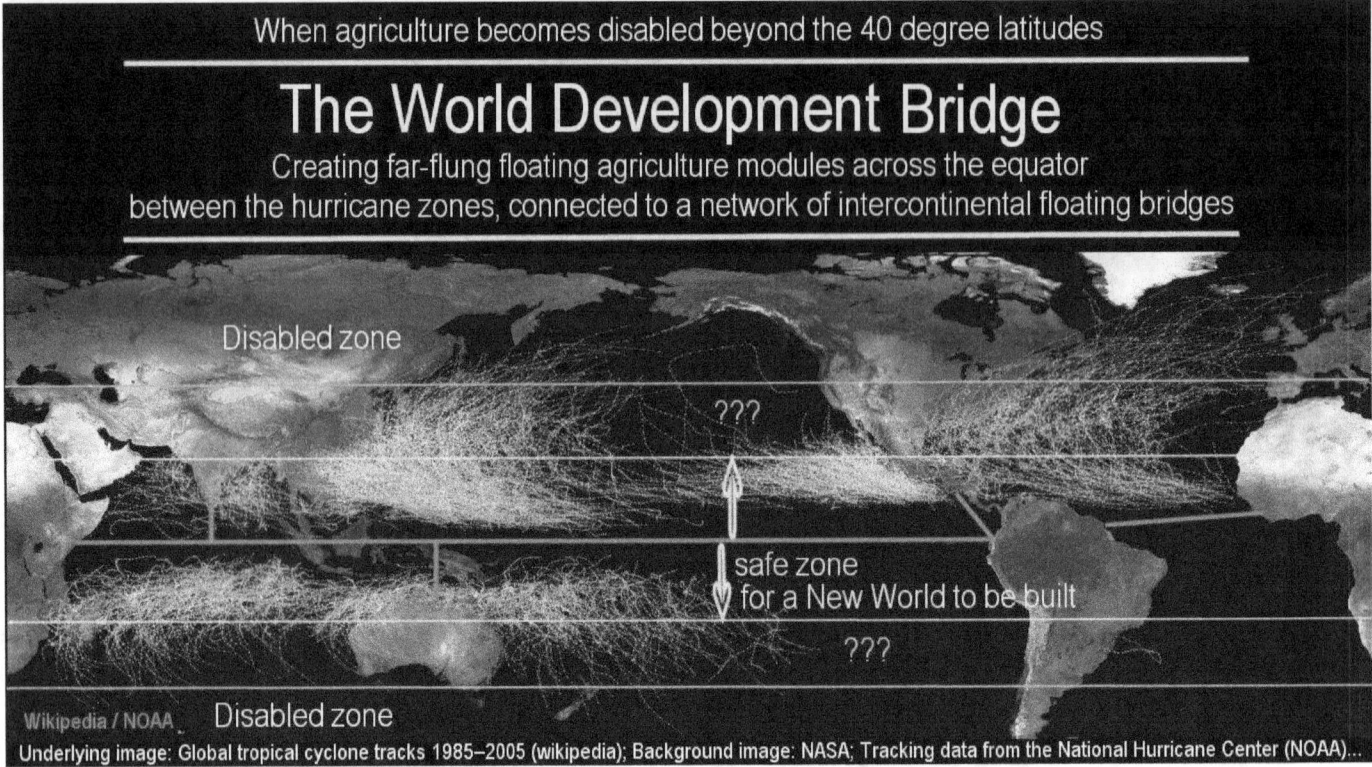

Only large-scale high-tech indoors agriculture has the capacity to nourish a society numbered in the billions, maybe ten or twenty billion strong.

That's why the people at Pinnacle Point survived

The people at Pinnacle Point had those amazing, down to Earth advantages over us, that we presently refuse to even consider. That's why the people at Pinnacle Point survived, while modern society assures itself, by its inaction, that it won't.

No provisions that would assure our continued living

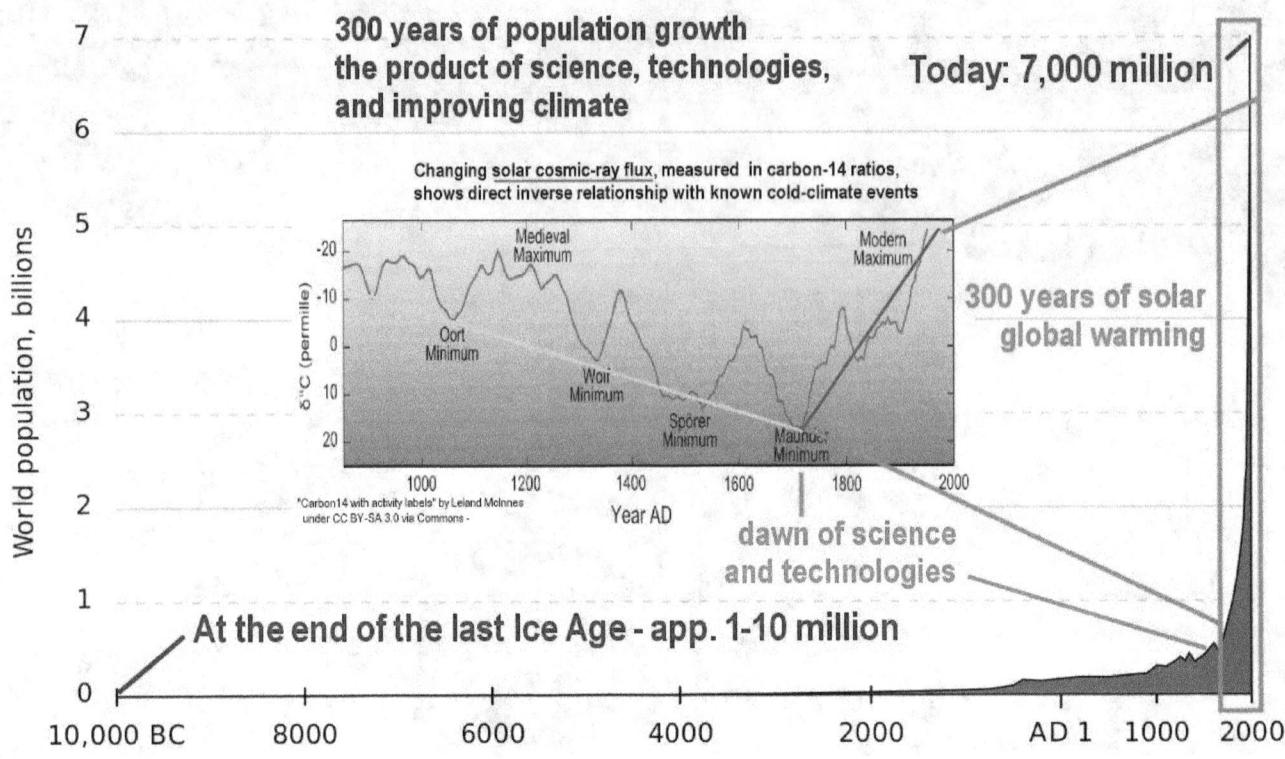

While no provisions are presently considered for any of the infrastructures to be built that would assure our continued living in the shadow of the already ongoing climate collapse, the corresponding population collapse is thereby actively invited to happen. This means that it will happen unless society is reversing is course. A middle-path course, is not possible.

But why is society so staunchly committed to its path to doom?

The culprit here, appears to be science fakery, a form of denial of the very truth that has been physically measured and is known. The political world is awash with such such types of fakery that deny reality and deny humanity.

Science has been especially targeted, and become trapped into fakery where nothing is actually real, by so-called experts and authorities, who in may ways proclaim that the real world doesn't actually exist, even while the opposite is obvious.

It is possible for humanity to step away from all forms of fakery, even from expert physicists who don't believe in physics anymore.

Humanity has the power to build for itself the New World with joy

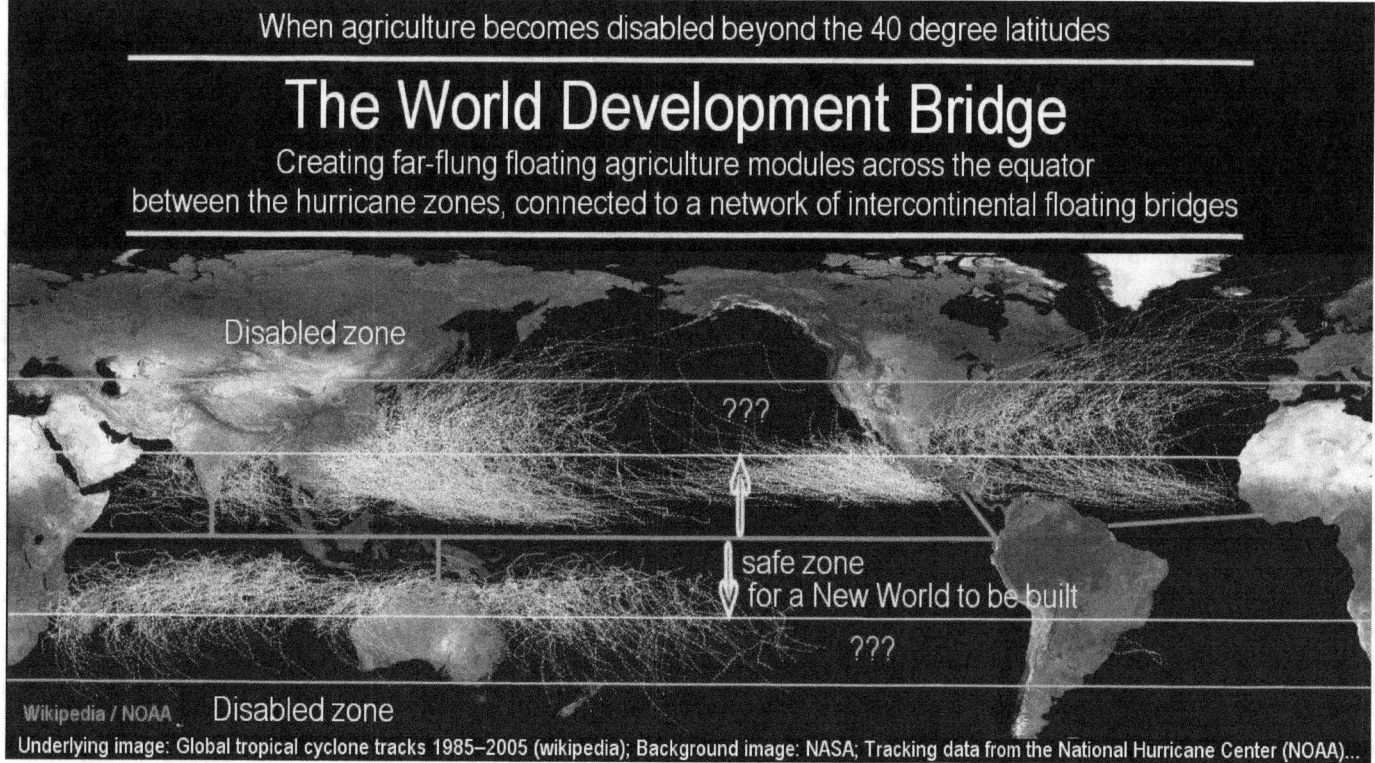

In moving forward from this dead end, humanity has the mental power to read the writing on the wall, and to find itself inspired to build for itself the New World in which the ongoing climate collapse cannot affect it, whereby the greatest-ever human population collapse can be avoided.

That's what science fakery denies. Science fakery is a death trap that would prevent humanity from having a future and trap it into a terrible doom. Society the world over, as human beings, has the capacity in full, to step away from this fakery and honor truth and life, with love for our humanity, and embrace reality with joy, and with power, and experience peace.

More Illustrated Science Books by Rolf A. F. Witzsche

www.ingramcontent.com/pod-product-compliance
Lightning Source LLC
Chambersburg PA
CBHW081023170526
45158CB00010B/3143